BRILLIANT PEOPLE, BIG IDEAS

TECHNOLOGY

Written by Rebecca Phillips-Bartlett

BookLife PUBLISHING

©2023
BookLife Publishing Ltd.
King's Lynn, Norfolk
PE30 4LS, UK

All rights reserved.
Printed in China.

A catalogue record for this book is available from the British Library.

HB ISBN: 978-1-80505-122-0
PB ISBN: 978-1-80505-381-1

Written by:
Rebecca Phillips-Bartlett

Edited by:
Kirsty Holmes

Designed by:
Isabella Croker

FSC
www.fsc.org
MIX
Paper from responsible sources
FSC® C113515

All facts, statistics, web addresses and URLs in this book were verified as valid and accurate at time of writing. No responsibility for any changes to external websites or references can be accepted by either the author or publisher.

Image Credits

All images are courtesy of Shutterstock.com, unless otherwise specified. With thanks to Getty Images, Thinkstock Photo and iStockphoto.

Recurring images – Andrew Rybalko, cosmaa, andromina, P-fotography, Mark Rademaker, vectorlab2D, Designsells. Cover – Andrew Rybalko, dimair, cosmaa, andromina, P-fotography, vectorlab2D, StrongPickles. 2–3 – VectorShow. 4–5 – Designsells, Ground Picture, Natalllenka.m. 6–7 – Jiang Zhongyan, Marish, Malika Keehl, Natalllenka.m, изображение_viber. 8–9 – Jan Schneckenhaus, Oxy_gen, StrongPickles, Morphart Creation. 10–11 – Procy, GoodStudio. 12–13 – Paul Maguire, Abscent Vector, Pogorelova Olga, Roi and Roi. 14–15 – Crisco 1492, GOLDMAN99, ST.art, Inactive design, vectorlab2D. 16–17 – Light show, WEB-DESIGN, beboy. 18–19 – svekloid, GOLDMAN99, robuart, Onica Alexandru Sergiu, Cuteness here. 20–21 – Sergio Andres Segovia, Vanatchanan, d1sk, notbad, Sudowoodo, Paulrclarke. 22–23 – MillaF, GoodStudio, Designsells.

Contents

Page 4	Big Ideas
Page 6	Li Tian
Page 8	Johannes Gutenberg
Page 10	Alexander Graham Bell
Page 12	Alice H. Parker
Page 14	Grace Hopper
Page 16	Hedy Lamarr
Page 18	Gerald 'Jerry' A. Lawson
Page 20	The Hall of Fame
Page 22	All You Need Is an Idea!
Page 24	Glossary and Index

Words that look like this can be found in the glossary on page 24.

Big Ideas

We can solve many human problems using science. This is called technology. Machines, computers and gadgets are all types of technology. We use technology so much that sometimes we can forget that everything had to be <u>invented</u> by someone.

Can you imagine life before we had television or even flushing toilets? Well, until some brilliant people had these big ideas, that is exactly how it was!

What technology do you use every day?

Li Tian

I am the Father of Fireworks. My name is Li Tian.

601–690

Fireworks

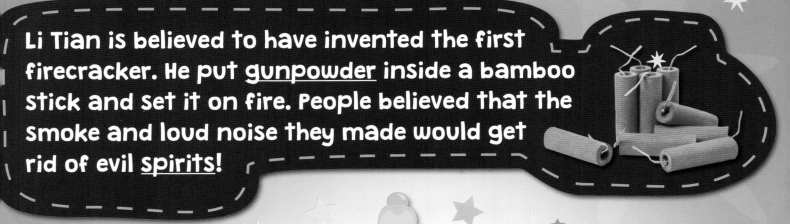

Li Tian is believed to have invented the first firecracker. He put <u>gunpowder</u> inside a bamboo stick and set it on fire. People believed that the smoke and loud noise they made would get rid of evil <u>spirits</u>!

When have you seen fireworks?

Today, we celebrate many special days with fireworks, such as Diwali, American Independence Day and the beginning of a new year.

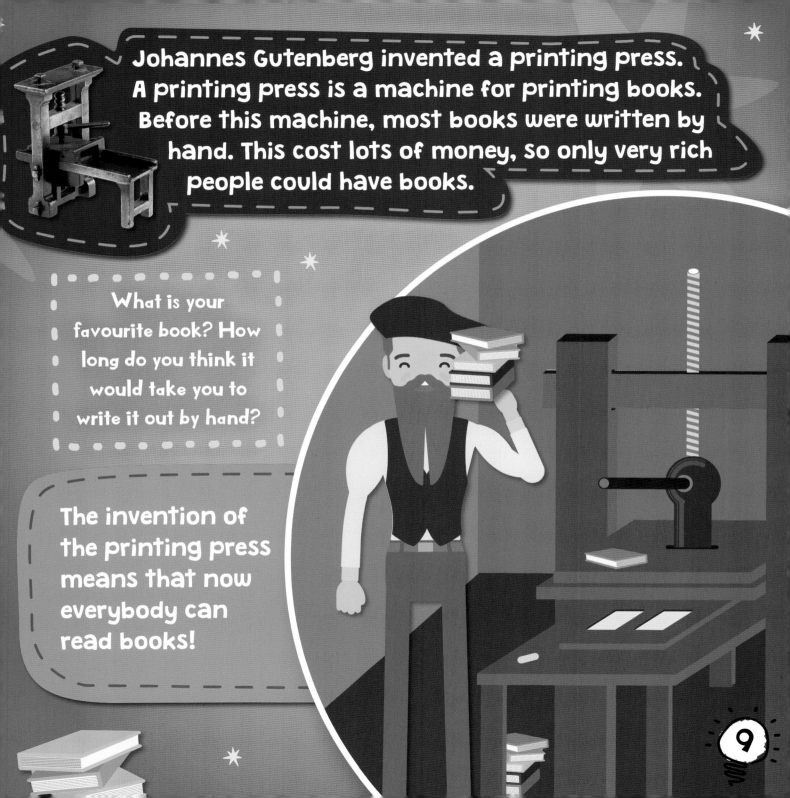

Johannes Gutenberg invented a printing press. A printing press is a machine for printing books. Before this machine, most books were written by hand. This cost lots of money, so only very rich people could have books.

What is your favourite book? How long do you think it would take you to write it out by hand?

The invention of the printing press means that now everybody can read books!

Alexander Graham Bell

I am Alexander Graham Bell. No one thought my invention would be possible!

1847–1922

Telephone

Just over 100 years ago, the only way to send messages far away was to write a letter. Alexander Graham Bell and Thomas Watson changed this. They invented a machine that could send a voice down a wire.

Soon, you could use this machine to send your voice across the world! Bell called this technology the 'electrical speech machine'. Today, we call it the phone.

Phones don't even need wires anymore!

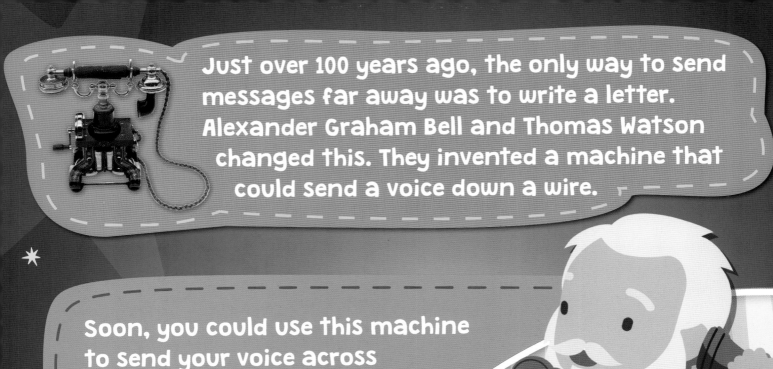

Alice H. Parker

1895–1920

I am Alice Parker. During winter, my house was way too cold!

Central Heating

12

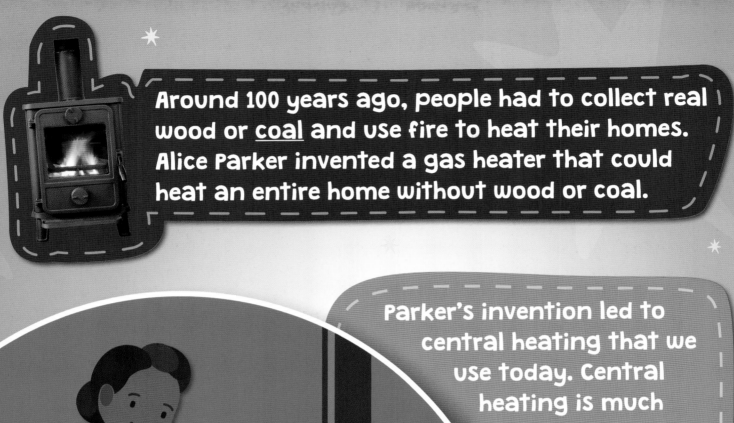

Around 100 years ago, people had to collect real wood or <u>coal</u> and use fire to heat their homes. Alice Parker invented a gas heater that could heat an entire home without wood or coal.

Parker's invention led to central heating that we use today. Central heating is much safer and easier than using a real fire!

Imagine having to go outside and find real wood every time you wanted to turn the heating up!

Grace Hopper

I am Grace Hopper. I invented a new language for computers!

1906–1992

Computer Programming

Early computers used numbers to know what to do. People who used them had to use a special computer language made up of numbers. Special computer languages are called code.

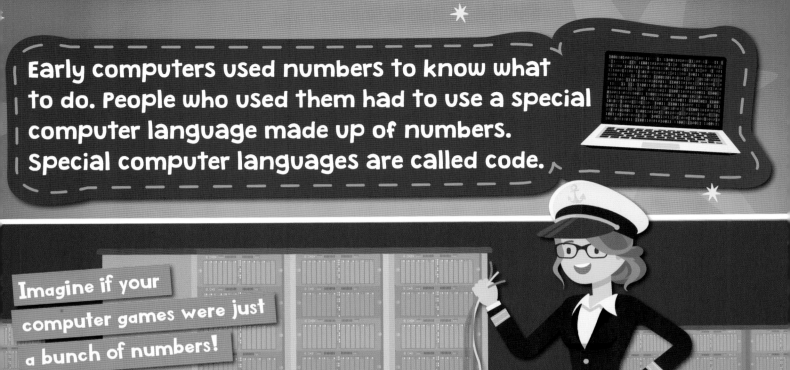

Imagine if your computer games were just a bunch of numbers!

Hopper invented a way to make computers understand words as well as numbers. This made it easier for more people to use computers.

Have you ever used Wi-Fi to look on the internet or play a game on a tablet? Or found the way somewhere using a Sat Nav? Have you ever sent a photo using Bluetooth?

What do you use **Wi-Fi** for?

None of this would have happened without Hedy Lamarr. She invented the technology that made all of this possible.

Lamarr is often called 'the mother of Wi-Fi'.

Gerald 'Jerry' A. Lawson

I am Jerry Lawson. My invention let people play videogames at home.

1940–2011

Videogame Cartridges

Early videogames had to be played on big machines. To play, you had to go to the videogame arcade where the machines were.

Arcade machine

Do you play videogames? What is your favourite game?

Jerry Lawson invented a way for people to play these games in their own homes. He invented games <u>cartridges</u>, which let you swap between different games on one console.

The Hall of Fame

Here are some more brilliant people who deserve a place in our Hall of Fame.

John Harrington

1561–1612

John Harrington invented the flushing toilet. Imagine the smell before he came up with this idea!

Sir Tim Berners-Lee

1955–now

Sir Tim Berners-Lee invented the World Wide Web, or the internet.

Ajay Bhatt

1957–now

USB cables are used to connect and charge many electronic <u>devices</u>. Ajay Bhatt invented USB to make electronics easier for more people to use.

All You Need is an Idea

From books to the internet, everything had to be invented. The people who came up with these inventions are brilliant, but it all started with their big ideas!

Glossary

actress	a woman whose job is acting on stage, in films or on television
cartridges	small cases that contain computer games and are put into a device to make it work
coal	a type of rock that can be burnt as fuel for a fire
devices	machines or inventions made to do something
gunpowder	a black powder which makes explosions
invented	when something new is made
spirits	beings that are not part of this world, such as ghosts and devils

Index

books 8–9, 22
central heating 12–13
computers 4, 14–15
fire 7, 13
fireworks 6–7
games consoles 19
internet 17, 21–22
letters 11
printing press 8–9
technology 4–5, 11, 17, 23
telephones 10
toilets 5, 20
USB 21
Wi-Fi 16–17